LEXICON LABS TITLE LIST

STEM
Python for Teens: A Step-by-Step Guide
QUANTUM COMPUTING for Smart Pre-Teens and Teens Ages 10-19
PHYSICS NERD: 1000+ Amazing And Mind-Blowing Facts About Physics
BIOLOGY NERD: 1000+ Amazing And Mind-Blowing Facts About Biology
CHEMISTRY NERD: 1000+ Amazing And Mind-Blowing Facts About Chemistry
ASTRONOMY NERD: 1000+ Amazing And Mind-Blowing Facts About Astronomy
AI for Smart Kids Ages 6-9: Discover How Artificial Intelligence is Changing the World
Code Breakers: A Practical Guide to Mastering Programming Languages and Algorithms
Quantum Nerd Quizmaster Edition: Quantum Quizzes that Educate, Entertain and Challenge
The AI Nerd: Quizmaster Edition Mind-Blowing AI Quizzes that Educate, Entertain and Challenge
AI for Smart Pre-Teens and Teens Ages 10-19: Using Artificial Intelligence to Learn, Think, and Create
(Spanish Translation) LA IA ESTÁ AQUÍ: Usa la Inteligencia Artificial para Aprender, Pensar y Crear

ENTREPRENEURSHIP
10 Life Hacks Every Teen Should Know
Innovation Handbook for Teen Entrepreneurs
Teen Innovators: 30 Teen Trailblazers and their Breakthrough Ideas

GREAT SCIENTISTS SERIES
Nikola Tesla: An Electrifying Genius
John von Neumann: The Giga Brain
Einstein: The Man, The Myth, The Legend
Newton: Genius of the Scientific Revolution
Darwin: Unlocking the Secrets of Evolution
Richard Feynman: The Adventures of a Curious Physicist
Marie Curie: Unleashing Radioactivity for Human Progress

GREAT INNOVATORS SERIES
Elon: A Modern Renaissance Man
Steve Jobs: The Visionary Innovator of Silicon Valley
Walt Disney: Creator of an Entertainment Empire

GREAT LEADERS SERIES
Cleopatra: Queen of the Nile
Gandhi: Freedom Fighter and Global Icon
Ben Franklin: Innovator, Statesman, Visionary
Churchill: The Spirit of an Indomitable Leader
St. Francis of Assisi: The Humble Servant of God
Alexander the Great: Conqueror, Visionary, Legend
Lincoln: Emancipator and Defender of the Union
George Washington: The First American President
Mark Antony: The Rise and Tragic Fall of a Roman Legend
Jefferson: Statesman, Visionary and the Third US President
Julius Caesar: The Rise and Fall of Rome's Greatest Leader
St. Ignatius of Loyola: Revolutionizing Faith and Education

GREAT EXPLORERS SERIES
Lewis and Clark: Blazing a Trail to the West
Magellan: First Circumnavigator of the Earth

Shackleton: Pioneering Explorer of the Antarctic
Columbus: The Explorer who Changed the World
Robert Falcon Scott: A Pioneer of Antarctic Exploration
Marco Polo: Intrepid Explorer who Bridged East and West
Captain Cook: The Legendary Seafarer, Navigator, and Explorer

GREAT ARTISTS SERIES

Taylor Swift: The Ascent of a Superstar
Andy Warhol: The Pop Art Phenomenon*
Van Gogh: Troubled Soul and Visionary Artist
Claude Monet: The Master of Impressionism
Michelangelo: Enigmatic Artist, Enigmatic Genius
Frida Kahlo: Unbroken Spirit: Artist, Activist, and Icon

TRIVIAL PURSUITS

Coffee Secrets
Vanishing Sun
Nobel Laureates 2023
Nobel Laureates 2024
Philosophy Brain-Teasers
Devilish Puzzles for Smart Kids
20th Century Wordsearch Extravaganza

COLORING BOOKS

Spaced Out
Vintage Auto
Orchid Fever
Serenity Now
Concept Cars
Mermaid Magic
Stoner Paradise
Swear.Laugh.Color.
Impossible Espresso
Perfectly Square Zen
Rebels of the Outback
Style with Confidence
Fantastical Creatures
Fabulous Fashionistas
National Parks Canada
101 Flower Arrangements
African Safari: A Mindfulness Coloring Book
Beautiful Australia: A Mindfulness Coloring Book
Magnificent South America: A Mindfulness Coloring Book
National Parks: Great Landscapes from America's Top National Parks

*Coming soon in Fall & Winter 2024

NOBEL LAUREATES 2024: The Winners and their Contributions Simply Explained

© 2024 by Dr. Leo Lexicon

Notice of Rights
All rights reserved. No portion of this publication may be reproduced, distributed, or transmitted in any form or by any means, electronic or mechanical, including photocopying, recording, or any information storage and retrieval system, without prior written permission from the author. Reproduction or translation of this work in any form beyond that permitted by Sections 107 or 108 of the 1976 United States Copyright Act is strictly prohibited. No companion books, summaries, or study guides are authorized under this notice. For other permission requests, please contact the author.

Liability Disclaimer
The information provided within this book is for general informational purposes only. While every effort has been made to keep the information up-to-date and correct, there are no representations or warranties, express or implied, about the completeness, accuracy, reliability, suitability, or availability with respect to the information, products, services, or related graphics contained in this book for any purpose. Any use of this information is at your own risk.

This book may contain affiliate links. In order to ensure trust and transparency, this disclosure complies with the rules set by the Federal Trade Commission. The support received through these affiliate links helps sustain the creation of valuable content. We appreciate the understanding and support of all readers.

The author, company, and publisher shall in no event be held liable for any loss or other damages, including but not limited to special, incidental, consequential, or other damages. This disclaimer applies to any damages caused by any failure of performance, error, omission, interruption, deletion, delay in transmission or transmission, defect, computer malware, communication line failure, theft, destruction, or unauthorized access to or use of records, whether for breach of contract, tort, negligence, or under any other cause of action. By reading this book, you agree to use the contents entirely at your own risk and that you are solely responsible for your use of the contents. The author, company, and publisher do not warrant the performance or applicability of any references listed in this book. All references and links are for informational purposes only and are not warranted for content, accuracy, or any other implied or explicit purpose

NOBEL LAUREATES 2024: The Winners and their Contributions Simply Explained

by

Dr. Leo Lexicon

NOBEL LAUREATES 2024: The Winners and their Contributions Simply Explained

Meet the 2024 Nobel Laureates!

"NOBEL LAUREATES 2024: The Winners and Their Contributions Simply Explained" provides a straightforward look at the individuals who were honored with the Nobel Prize in 2024. Their contributions are detailed in a clear and engaging manner. This book takes you through the diverse fields of Physics, Chemistry, Medicine, Literature, Peace, and Economic Sciences, and describes how the winner(s) in each category made strides in their respective domains.

The 2024 laureates have pushed the boundaries of knowledge across multiple fields, reshaping our understanding of health, artificial intelligence, chemistry, literature, peace, and economics. Their achievements range from revolutionary medical discoveries that promise to benefit millions, to breakthroughs in neural networks that have propelled advancements in artificial intelligence, transforming how we interact with technology. In chemistry, the integration of experimental and computational approaches has brought about new possibilities for life-saving medical therapies. In literature, a profound exploration of trauma and identity has captured the human experience in an unforgettable way, inspiring new generations. The Peace Prize was awarded to those advocating for nuclear disarmament and giving a voice to survivors, bringing renewed hope and fostering global peace. Finally, the laureates in economics have illuminated the vital role institutions play in shaping economic outcomes and prosperity. Each of these achievements represents a significant leap forward, showcasing the resilience, dedication, and innovative spirit of individuals committed to improving our world.

Like its predecessor in 2023, this book is a reliable source for those interested in the Nobel Prize, and created in order to offer a clear view into the remarkable achievements recognized in 2024. Whether you are a scientist, a student, a lover of literature, or simply curious about the world, this book will leave you inspired by the remarkable individuals who have dedicated their lives to pushing the boundaries of knowledge and fighting for a better future.

What are you waiting for? It is time to dig in!

Dr. Leo Lexicon is an educator and author. He is the founder of Lexicon Labs, a publishing imprint that is focused on creating entertaining and educational books for active minds.

CONTENTS

Chapter 1: A Brief History 1
    The Origins of the Nobel Prize 1
    Anticipation and Significance 2
    Major Landmarks in Nobel History 3
    Impact on Science and Human Achievement 4
    Amazing Facts 6

Chapter 2: Physiology or Medicine 9
    Detailed Explanation 9
    Historical Context 10
    The Breakthrough 11
    Personal Journeys of the Laureates 12
    Impact and Legacy 12
    Impact and Legacy in Medicine and Research 13
    Fun Facts about Previous Laureates in Physiology 14

Chapter 3: Physics 16
    Nobel Prize in Physics 2024 16
    Detailed Explanation 16
    The Foundations of Neural Networks 17
    Geoffrey Hinton and the Backpropagation Algorithm 18
    John Hopfield and Associative Memory 19
    Historical Context 19
    The Breakthroughs in Deep Learning 20
    Personal Journeys of the Laureates 21
    Inspiring the Leading Lights of AI 22
    Longer-term Impact 24

Fun Facts About Previous Laureates in Physics 24

Chapter 4: Chemistry 26

  Nobel Prize in Chemistry 2024 26
  Detailed Explanation 26
  Transforming Protein Science with AI 27
  The Foundations of Protein Research 27
  From Experimental to Computational Approaches 30
  Real-World Applications and Impact 31
  Personal Journeys of the Laureates 31
  Inspiring Future Generations 32
  Long-Term Impact 32
  Fun Facts About Previous Laureates in Chemistry 32

Chapter 5: Literature 34

  Nobel Prize in Literature 2023 34
  Detailed Explanation 34
  The Foundations of Han Kang's Literary Works 35
  Han Kang's Exploration of Trauma and Identity 36
  Historical and Social Context 37
  Personal Journey 38
  Impact and Legacy 39
  Inspiring a New Generation 39
  Conclusion 40
  Fun Facts About Previous Laureates in Literature 40

Chapter 6: Peace 42

  Nobel Peace Prize 2024 42
  Detailed Explanation 42
  A Voice for the Survivors 43

- Advocating for Nuclear Disarmament 44
- Survivors as Symbols of Resilience 44
- Building Bridges for Global Peace 45
- The Path to Disarmament 45
- Fun Facts About Previous Laureates in Peace 46

Chapter 7: Economics 48

- Nobel Peace Prize in Economic Sciences 2024 48
- Detailed Explanation 48
- The Foundations of Institutional Economics 49
- Why Institutions Matter? 50
- Effects of Institutional Change 51
- The Role of Political Economy 51
- Historical Context 52
- Understanding Institutions and Prosperity 53
- Personal Journeys of the Laureates 53
- Inspiring Future Economists 54
- Longer-term Impact 55
- Fun Facts About Previous Laureates in Economics 56

FREE SAMPLE: 60

Chapter 3 60

General Relativity 60

- Understanding Spacetime 60
- Key Concepts Explained 63
- Impact on Physics 66
- Controversies and Acceptance 68
- A Final Word 71

Chapter 1: A Brief History

The Origins of the Nobel Prize

The origin of the Nobel Prize is rooted in the complex personality of Alfred Nobel, a Swedish chemist, engineer, and inventor who amassed a vast fortune from his 355 patents. Nobel's most famous invention, dynamite, revolutionized the construction and mining industries. However, the potential for its use in warfare led to a deep sense of guilt for Nobel, culminating in an event that would transform his legacy. In 1888, a French newspaper published a premature obituary titled *"The Merchant of Death is Dead,"* referring to Nobel. This obituary, intended for his brother, shook Nobel to his core, making him reflect on how his life would be remembered.

Nobel did not wish to be remembered for his inventions alone—especially those with destructive potential. In 1895, less than a year before his death, he signed his will at the Swedish-Norwegian Club in Paris. This will dedicated the majority of his fortune to establishing five prizes in the fields of Physics, Chemistry, Medicine, Literature, and Peace, each awarded to individuals or organizations who had provided the "greatest benefit to humanity." In 1968, the Sveriges Riksbank established the Prize in Economic Sciences in Memory of Alfred Nobel, broadening the reach of the awards.

Despite skepticism from Nobel's family and peers about the practicality of his vision, the first Nobel Prizes were awarded in 1901. Over time, the Nobel Prizes grew into the most prestigious awards globally, symbolizing human achievement, discovery, and dedication to improving the world.

Anticipation and Significance

As the calendar approaches October each year, the scientific and academic communities, alongside global citizens, eagerly await the announcement of the Nobel Prizes. This year, 2024, has been no exception! These awards, based on over a century of tradition, signal not only the recognition of individual achievements but also the celebration of intellectual and cultural advancement. For many laureates, the prize is a culmination of years, if not decades, of work, perseverance, and sacrifice. For the world, it serves as a reminder of the power of human ingenuity, perseverance, and the tireless pursuit of knowledge, demonstrating what can be achieved when people dedicate themselves to pushing the boundaries of what is possible.

Fig. Albert Einstein, Nobel Laureate in Physics, 1921

The announcement ceremonies, held in Stockholm and Oslo, are a unique convergence of global intellectual attention, bringing together people from all walks of life to celebrate the remarkable achievements that have made a difference in the world. Across disciplines, from the foundational sciences to literature and humanitarian efforts, the Nobel Prizes celebrate the relentless drive to push boundaries and the spirit of exploration that propels humanity forward. Each laureate represents a story of dedication, breakthroughs, and often collaboration, reflecting Nobel's vision of international recognition that transcends national borders, politics, and cultures, fostering a sense of unity and shared progress.

The significance of these awards extends far beyond the personal recognition of the laureates. These awards galvanize fields, drawing attention to breakthrough ideas that may otherwise remain obscure, and sparking new avenues of research and discussion. They inspire future generations of scholars, scientists, writers, and activists, highlighting the potential for their efforts to be recognized on a global stage. Moreover, the prizes serve as a beacon of hope, reminding society of the value of investing in research, creativity, and advocacy, and the impact that such investments can have on shaping a better future for all.

Major Landmarks in Nobel History

The Nobel Prizes have continually shaped the narrative of human progress, with landmark moments that have resonated throughout history. Some of the most notable milestones include:

- **1903:** Marie Curie becomes the first woman to receive a Nobel Prize, sharing the Physics prize with her husband Pierre Curie and Henri Becquerel for their work on radioactivity. In 1911, she wins again in Chemistry, solidifying her place as one of the greatest scientists in history.
- **1921:** Albert Einstein, renowned for his contributions to theoretical physics, is awarded the Nobel Prize in Physics, specifically for his work on the photoelectric effect—a discovery that formed the foundation for quantum theory.

- **1964:** Martin Luther King Jr. is awarded the Nobel Peace Prize for his nonviolent resistance against racial segregation in the United States. His recognition marked a key moment in the global civil rights movement and reinforced the connection between peace efforts and social justice.
- **1979:** Mother Teresa is honored with the Nobel Peace Prize for her humanitarian work in India. She donated her prize money to the poor in Calcutta, exemplifying the spirit of the Peace Prize.
- **2004:** Wangari Maathai, a Kenyan environmental activist, becomes the first African woman to receive the Nobel Peace Prize. Her work in sustainable development, democracy, and peace through the Green Belt Movement had a profound impact on African environmental policies.

The Nobel Prizes are also intertwined with moments of political and social significance. For instance, when Liu Xiaobo, a Chinese dissident, was awarded the Nobel Peace Prize in 2010, it highlighted issues of human rights and government repression. Such moments underscore the power of the Nobel Prizes not only as scientific and cultural recognitions but as instruments of global discourse on peace, justice, and freedom.

Impact on Science and Human Achievement

The Nobel Prize has left an indelible mark on human achievement, particularly in the fields of science, literature, and humanitarian efforts. The accolades shine a light on the remarkable contributions made by individuals and teams that shape the world as we know it.

In science, Nobel Prizes have been key in promoting and popularizing groundbreaking discoveries. From the structure of DNA to the Higgs boson particle, many scientific advances that have redefined human understanding have been recognized through these awards. Beyond the individual recognition, the prize often provides crucial momentum to entire fields of study. For example, the discovery of CRISPR technology, which won the Chemistry prize in 2020, has propelled genetic research into new territories with possibilities for revolutionizing medicine.

In literature, the Nobel Prize often elevates authors whose works reflect the diverse and complex narratives of the human experience. From Toni Morrison to Kazuo Ishiguro, the laureates have offered timeless stories and social critiques, many of which have gained broader readership following their Nobel win. The recognition also highlights the importance of literature in preserving cultural heritage and advancing social awareness.

The Peace Prize has played a unique role in influencing global politics. It has brought attention to issues such as nuclear disarmament, climate change, and human rights, influencing international policies and prompting diplomatic efforts. Figures like Nelson Mandela and organizations such as the International Campaign to Ban Landmines serve as examples of how the Nobel Peace Prize can be a catalyst for change.

One of the most profound impacts of the Nobel Prizes is the creation of role models. Laureates like Malala Yousafzai, the youngest-ever Nobel Peace Prize recipient, who advocated for girls' education, have inspired millions to advocate for global causes. Similarly, scientists like Emmanuelle Charpentier and Jennifer Doudna, who pioneered CRISPR, are now household names encouraging young women to enter STEM fields.

The Nobel Prizes also amplify stories of perseverance. For example, Barbara McClintock, whose groundbreaking research in genetics was not initially understood or accepted, was awarded the Nobel Prize in 1983 for her discovery of "jumping genes." Her story is a testament to the importance of resilience in scientific inquiry.

Amazing Facts

Now, here are some intriguing facts about the Nobel Prizes that continue to fascinate and surprise:

- **Oldest Winner:** John B. Goodenough became the oldest Nobel laureate at the age of 97 when he was awarded the 2019 Nobel Prize in Chemistry for his work on lithium-ion batteries.
- **Largest Group Awarded:** In 2007, the Nobel Peace Prize was awarded to the Intergovernmental Panel on Climate Change (IPCC) and Al Gore. The IPCC consisted of more than 2,000 scientists from across the globe, making it one of the largest groups to receive the prize.
- **Dual Family Laureates:** Beyond the Curies, the Nobel Prize has seen several families receive multiple awards. For example, the Braggs, a father-son duo, both received the Nobel Prize in Physics in 1915 for their work in X-ray crystallography.
- **First Organization to Win Twice:** The United Nations High Commissioner for Refugees (UNHCR) has won the Nobel Peace Prize twice, in 1954 and 1981, in recognition of its work with refugees around the world.
- **Unlikely Nobelists:** In 1945, the Nobel Prize in Physics was awarded to Wolfgang Pauli, an Austrian theoretical physicist, who famously didn't conduct any experiments himself but contributed to the theory of quantum mechanics.
- **Remarkable Resignation:** Boris Pasternak, a Russian novelist awarded the Nobel Prize in Literature in 1958, was forced by the Soviet government to decline the prize for his novel *Doctor Zhivago*.
- **First Posthumous Nomination Controversy:** While Nobel Prizes cannot be awarded posthumously, a notable exception nearly occurred in 1961. Swedish diplomat Dag Hammarskjöld was posthumously awarded the Nobel Peace Prize after his untimely death in a plane crash, as his nomination had been submitted before his death.

- **Nobel Prize in Mathematics?** Despite the diverse range of disciplines covered by the Nobel Prizes, there is no Nobel Prize in Mathematics. Rumor has it that Nobel intentionally excluded mathematics, possibly due to personal grudges, although there is no concrete evidence to support this.
- **The Missing Laureate:** In 1939, the Nobel Committee selected Carl von Ossietzky, a German journalist, as the Peace Prize laureate for his anti-Nazi activism. However, Adolf Hitler forbade him from attending the ceremony, and Ossietzky remained in Nazi detention until his death, never able to claim his award.
- **Only One Nobel for Environmental Efforts:** Despite the growing global focus on environmental concerns, the only environmentalist to have won the Nobel Peace Prize to date is Wangari Maathai (2004) for her efforts in sustainable development and reforestation.
- **Laureate in Exile:** Aleksandr Solzhenitsyn, awarded the Nobel Prize in Literature in 1970, was unable to accept his award in person due to threats from the Soviet government. He did not retrieve it until 1974, after his exile from the Soviet Union.
- **First Music-Related Nobel:** In 2016, Bob Dylan became the first songwriter to receive the Nobel Prize in Literature. Although his win sparked a debate about whether song lyrics qualify as literature, Dylan's recognition opened the door for new ways to view creative works.
- **Most Nominations Without a Win:** Mahatma Gandhi was nominated for the Nobel Peace Prize five times but never won. The Nobel Committee publicly expressed regret for not awarding him the prize before his assassination in 1948.
- **Prize for a Single Word:** Shirin Ebadi, the 2003 Nobel Peace Prize winner from Iran, was specifically recognized for her efforts in promoting democracy and human rights. One word from her acceptance speech—"dialogue"—resonated globally, becoming the theme of numerous international conferences on peace that followed.
- **Nobel Foundation's Self-Correcting Decision:** In 1948, no Nobel Peace Prize was awarded as the Committee found "no suitable living candidate." The decision is often viewed

in hindsight as an acknowledgment of Gandhi's missed opportunity, though no prize can be awarded retrospectively.
- **The Technical "Nobel" Prize:** The *Nobel Prize in Economic Sciences* (officially called the Sveriges Riksbank Prize in Economic Sciences in Memory of Alfred Nobel) was not part of Alfred Nobel's original will. It was added in 1968 by the Swedish central bank, making it the only prize not directly created by Nobel himself.

That concludes our brief introduction to the Nobel Prizes. Let us now move on to the main course, the 2024 Prizes and the winners themselves!

Chapter 2: Physiology or Medicine

Category: Physiology or Medicine

Winners: Victor Ambros and Gary Ruvkun

Associated Institutions: University of Massachusetts Medical School (Ambros), Harvard Medical School (Ruvkun)

Summary: Ambros and Ruvkun were awarded the 2024 Nobel Prize in Physiology or Medicine for their groundbreaking discovery of microRNA (miRNA), a class of small RNA molecules that regulate gene expression. This discovery revealed a new post-transcriptional gene regulatory mechanism, essential for multicellular organisms' development and function. Their findings have profoundly impacted biological research and medicine.

Detailed Explanation

The Nobel Prize in Physiology or Medicine for 2024 recognized the unique contributions of Victor Ambros and Gary Ruvkun in uncovering the role of microRNA (miRNA) in gene regulation. Their research introduced a new understanding of how genes are regulated beyond the transcriptional level. MicroRNAs control the translation of messenger RNA (mRNA), fine-tuning the production of proteins, which are vital to a variety of biological processes, from cell differentiation to immune response. This discovery has reshaped molecular biology and has become a cornerstone in studying genetic regulation and related diseases.

Historical Context

Before the discovery of miRNAs, scientists thought that genes were mostly controlled by transcription factors. These factors either turned genes on or off by controlling how they made mRNA, which is like a message that tells the cell what protein to make. Once the mRNA was made, it was assumed that it would always be used to make a protein. Ambros and Ruvkun's research showed there was another level of control—one that happens after the mRNA is made but before it turns into a protein.

Ambros and Ruvkun started their research by studying the development of a small worm called *Caenorhabditis elegans*, which is often used in genetics research. In the early 1990s, they found lin-4, the first known miRNA. They showed that this small RNA molecule, only 22 building blocks long, could attach to the mRNA of another gene called lin-14. By doing this, lin-4 could stop lin-14 from being turned into a protein, which helped control the timing of the worm's development. This was the first time scientists realized that an RNA molecule that does not make a protein could still have an important job.

Later on, in 2000, Ruvkun discovered another important miRNA called let-7. They found that let-7 was present in many different animals, including humans, which showed that miRNA-based regulation was an important mechanism for all complex living things.

These findings were counterintuitive and surprising as they introduced an entirely new layer of gene regulation that scientists had not previously considered. The conventional wisdom was that once mRNA was created from DNA through transcription, the gene's role was largely finished, with mRNA consistently being translated into proteins. Transcription factors were understood as the primary control mechanism, turning genes on or off by managing mRNA production. However, Ambros and Ruvkun's work revealed that mRNAs could control gene expression at a post-transcriptional level, meaning they could bind to mRNA and prevent it from being translated into a protein, even after the mRNA was made. This was surprising because it contradicted the existing assumption that mRNA would always lead to protein production. The fact that these small RNA molecules, which do not code for proteins themselves, could still have such a significant regulatory role, was a significant breakthrough. It essentially showed that genetic control was more complex and nuanced than previously thought, introducing a new dimension to our understanding of biology.

The Breakthrough

Ambros and Ruvkun's work was initially met with skepticism, as the role of small, non-coding RNAs in gene regulation was unprecedented at the time. However, their perseverance paid off as subsequent research confirmed the existence of hundreds of miRNAs across a wide range of species. These small RNA molecules are now known to regulate the expression of nearly one-third of human genes, highlighting their importance in both normal biological functions and disease.

Their discovery fundamentally altered the understanding of gene regulation. Rather than being a one-way street from DNA to mRNA to proteins, Ambros and Ruvkun revealed a regulatory loop, where miRNAs could silence or reduce the production of proteins by binding to mRNAs and preventing their translation. This provided cells with a powerful tool to modulate gene expression in response to developmental cues and environmental conditions.

Over time, research revealed that miRNAs are involved in virtually every aspect of biology, from early embryonic development to immune response and metabolism. Dysregulation of miRNAs has been linked to a variety of diseases, including cancer, cardiovascular diseases, and neurological disorders. Their precise role in regulating gene expression makes them attractive targets for therapeutic intervention, with several clinical trials underway to explore the potential of miRNA-based treatments.

Personal Journeys of the Laureates

Both Victor Ambros and Gary Ruvkun encountered significant challenges on their path to this groundbreaking discovery. Early in their careers, their research was viewed with skepticism, and securing funding for such unconventional ideas was difficult. Despite these obstacles, they persisted, driven by their curiosity and belief in the potential of small RNA molecules to regulate gene expression.

Ambros and Ruvkun worked in separate labs, but their collaboration was crucial to the development of the miRNA field. Their independent discoveries of lin-4 and let-7 were complementary, and they frequently shared insights and data that accelerated the pace of research in RNA biology. Ambros focused on the functional aspects of miRNAs, while Ruvkun investigated their evolutionary conservation and molecular mechanisms. Together, their work laid the foundation for what would become one of the most rapidly growing fields in molecular biology.

Impact and Legacy

The discovery of miRNAs has had an immense impact on biology and medicine. It has provided crucial insights into how cells regulate gene expression and has opened up new avenues for understanding development and disease. The role of miRNAs in cancer, in particular, has attracted significant attention. Some miRNAs function as oncogenes, promoting cancer development by downregulating tumor-suppressor genes, while others act as tumor suppressors themselves. Understanding these roles has led to the exploration of miRNAs as both diagnostic biomarkers and therapeutic targets in cancer treatment.

Beyond cancer, miRNAs are also being investigated for their roles in cardiovascular diseases, metabolic disorders, and neurodegenerative diseases. Their ability to fine-tune gene expression makes them ideal candidates for therapies that require precise regulation of protein production. miRNA-based therapies, which aim to either inhibit or replace specific miRNAs, are currently being tested in clinical trials for conditions that have proven difficult to treat with conventional therapies.
The discovery of miRNAs has also had implications beyond human health. In agriculture, miRNAs play a role in regulating plant growth and development, influencing crop yield and stress resistance. Researchers are exploring the potential to manipulate miRNA pathways to improve crop resilience and productivity in the face of climate change.

Ambros and Ruvkun's work has fundamentally altered the landscape of molecular biology. What began as a study of developmental timing in nematodes has grown into a vast field of research that spans multiple disciplines. Their discovery of miRNAs has not only provided answers to long-standing questions about gene regulation but has also opened new frontiers in the treatment of disease and the improvement of agricultural practices. As miRNA research continues to advance, its applications in medicine and biotechnology are expected to expand further. Ambros and Ruvkun's work has laid the foundation for a new era of RNA biology, and their discovery of miRNAs will continue to shape scientific research for decades to come.

Impact and Legacy in Medicine and Research

The discovery of miRNAs has broad implications for future medical diagnostics and treatments. Today, miRNAs are already being used in clinics to help diagnose certain cancers and to predict how patients will respond to different treatments. Researchers are working on new drugs that target miRNAs, and these therapies look especially promising in treating cancers and heart diseases. In the future, miRNA research could lead to even more precise treatments. Because miRNAs can control how genes are expressed, they are ideal for targeting specific diseases at the genetic level, potentially leading to personalized medicine where treatments are tailored to each individual's genetic makeup. For example, miRNA-based treatments could help turn off harmful genes in cancer or help restore normal gene function in genetic disorders.

The future of miRNA research is also tied closely to advancements in technology. With better tools to detect and manipulate miRNAs, scientists can create more targeted drugs and diagnostics. This precision could significantly reduce side effects and improve the effectiveness of treatments. Furthermore, ongoing research suggests that miRNAs might play a role in treating neurological disorders, metabolic diseases, and even infectious diseases, offering a wide range of potential medical applications.

Overall, miRNAs are shaping up to be a powerful tool in modern medicine. Their ability to fine-tune gene expression gives them a unique edge in treating complex diseases, and with new technologies emerging, their impact on personalized and precise healthcare is likely to grow.

Fun Facts about Previous Laureates in Physiology

- **Frederick Banting (1923)**: Frederick Banting, who was awarded for the discovery of insulin, was the youngest ever Nobel laureate in Physiology or Medicine, receiving the prize at age 32.

- **Albert Schweitzer (1952)**: Schweitzer was not only a medical doctor but also a renowned philosopher, theologian, and musician. His prize was awarded for his humanitarian work, making him unique among medicine laureates.

- **Barbara McClintock (1983)**: McClintock, awarded for her discovery of transposable genetic elements, was the first woman to win the Nobel Prize in Physiology or Medicine unshared.

- **Luc Montagnier and Françoise Barré-Sinoussi (2008)**: They were recognized for discovering the HIV virus, contributing greatly to the understanding and treatment of AIDS.

- **Tu Youyou (2015)**: Tu Youyou discovered artemisinin, a drug used to combat malaria, drawing from traditional Chinese medicine. She was the first Chinese citizen to win the Nobel Prize in Medicine.

Chapter 3: Physics

Nobel Prize in Physics 2024

Category: Physics

Winners: Geoffrey Hinton and John Hopfield

Associated Institutions: University of Toronto (Hinton), Princeton University (Hopfield)

Summary: Hopfield and Hinton received the 2024 Nobel Prize in Physics for their pioneering work in applying statistical physics concepts to develop artificial neural networks. Their research has been fundamental in advancing machine learning and artificial intelligence, enabling machines to recognize patterns in large datasets. Their work has applications across various fields, including particle physics, material science, and astrophysics, as well as in everyday technologies like facial recognition and language translation

Detailed Explanation

The 2024 Nobel Prize in Physics was awarded to two pioneers in artificial neural networks, Geoffrey Hinton and John Hopfield. Their revolutionary contributions have laid the foundation for deep learning, a technology that has reshaped not just artificial intelligence (AI) but industries and academic disciplines around the globe. Hinton and Hopfield's research unlocked the potential for machines to mimic human cognitive processes, a feat once thought to belong purely to the realm of science fiction. This chapter will examine their discoveries, the historical context that preceded their breakthroughs, and the profound impact their work continues to have.

When the announcement was first made, a lot of people (especially on social media) wondered about the connection between physics and developments in computer science and artificial intelligence. It turns out that these phenomena are deeply rooted in physics.

Neural networks, which emulate the processes of the human brain, are not only computational models but also involve principles from physics, such as energy minimization and optimization. Hopfield, in particular, applied concepts from statistical mechanics, like atomic spins in materials, to develop the Hopfield network, a type of associative memory system. Hinton's backpropagation algorithm, although rooted in computer science, also relies on mathematical techniques common in physics. The committee recognized their work as physics because it deals with fundamental questions about how systems—whether they are neurons or atoms—process and store information. The application of these principles to neural networks has revolutionized artificial intelligence, influencing a wide range of scientific and practical domains.

The Foundations of Neural Networks

Artificial neural networks (ANNs), modeled after biological neurons, are mathematical constructs designed to simulate the way the human brain processes information. In their early stages, these models struggled to live up to their potential due to technical limitations such as inefficient training methods, the vanishing gradient problem, and a lack of understanding of how to store and retrieve information in a neural system. The contributions of

Geoffrey Hinton and John Hopfield offered solutions to these challenges and paved the way for the explosive growth of AI.

Geoffrey Hinton and the Backpropagation Algorithm

Geoffrey Hinton, often referred to as the "godfather of deep learning," introduced backpropagation in the mid-1980s, a method for training artificial neural networks. This algorithm allowed machines to adjust weights and biases in multi-layer networks, making it possible for deep neural networks to "learn" from data. Without backpropagation, it would have been practically impossible to train the deep networks that are responsible for modern breakthroughs in image recognition, natural language processing, and autonomous decision-making.

Prior to Hinton's work, training deep networks faced several obstacles, especially with multi-layer networks. Early neural networks were limited to a few layers because training deeper layers was computationally infeasible. The vanishing gradient problem—where gradients became increasingly smaller as they were propagated back through the layers—meant that learning in deeper networks was prohibitively slow and ineffective. Hinton's backpropagation algorithm elegantly addressed this issue, enabling the networks to distribute errors efficiently during the learning process and, thus, make meaningful adjustments across layers.

Hinton's development of backpropagation transformed neural networks from theoretical constructs into practical tools that could handle real-world data at scale. By allowing machines to fine-tune their internal parameters through iterative learning, Hinton set the stage for modern AI systems to excel in complex tasks such as recognizing faces in images, translating languages, and even generating human-like text.

John Hopfield and Associative Memory

John Hopfield, on the other hand, is best known for his development of the Hopfield network, a form of recurrent artificial neural network that acts as an associative memory system. Introduced in the early 1980s, Hopfield's model allowed neural networks to store patterns and retrieve them based on partial or noisy input. This concept has been especially influential in both AI and neuroscience, offering insight into how biological brains might store and recall information.

Hopfield's associative memory model works by minimizing an energy function, akin to how biological neural networks seem to "settle" into stable states that represent memories. This energy minimization process allows the Hopfield network to correct noisy or incomplete inputs, effectively reconstructing the original memory. The implications of this for AI were profound, as it allowed machines to handle ambiguous data more robustly and mimic human-like recall.

Hopfield's work was particularly notable for bridging the gap between neuroscience and artificial intelligence. His insights into how biological neurons might behave as networks capable of solving optimization problems inspired new ways of thinking about both AI algorithms and cognitive processes in humans. His contributions, like Hinton's, helped lay the groundwork for deep learning technologies by showing that neural networks could perform not just computation but also memory storage and retrieval.

Historical Context

Before the contributions of Hinton and Hopfield, neural networks were not taken seriously by much of the scientific community. Early work on neural networks, such as the Perceptron model developed by Frank Rosenblatt in the 1950s, showed some promise, but its limitations became apparent by the 1970s. The inability of these early models to solve non-linear problems led to widespread skepticism about the potential of neural networks, and for several decades, research in this area stalled.

During the so-called "AI winter" of the 1970s and 1980s, funding and interest in AI research waned as it became clear that existing neural network models were not capable of tackling complex tasks. However, researchers like Hinton and Hopfield remained undeterred. Their persistence and belief in the potential of neural networks, even when the field had fallen out of favor, ultimately resulted in the breakthroughs that reignited the AI revolution in the 21st century.

It is important to recognize that the development of deep learning owes much to advances in computational power and the availability of large datasets. The theoretical work laid down by Hinton and Hopfield in the 1980s and 1990s became practical when coupled with the exponential growth in computing capacity and data storage in the early 2000s. With the introduction of graphical processing units (GPUs) for parallel processing, deep learning networks could be trained on massive datasets, enabling the breakthroughs we see today in fields ranging from healthcare to autonomous driving.

The Breakthroughs in Deep Learning

Hinton's backpropagation algorithm enabled the training of deep neural networks, a key milestone in AI. The architecture and algorithms Hinton developed in the 1980s are the backbone of most deep learning applications today. In 2012, Hinton's deep learning methods led to a landmark achievement in AI when his team's neural network outperformed traditional methods in the ImageNet competition, a benchmark for image classification tasks. This achievement marked the dawn of the AI era, where deep learning became the dominant paradigm for solving complex, data-driven problems.

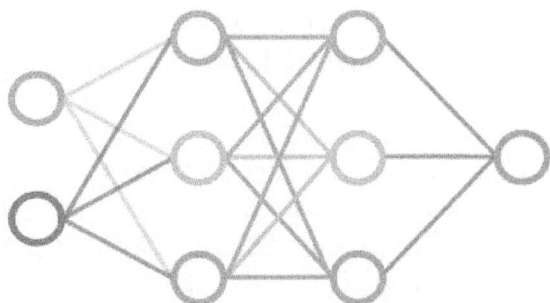

Fig. Backpropagation in Neural Networks

John Hopfield's contributions are equally transformative. His associative memory networks introduced a new way of thinking about how systems could manage noise, ambiguity, and partial data. In practical terms, this meant that AI systems inspired by Hopfield's work could function more reliably in real-world environments where data is rarely perfect. For example, in medical imaging, Hopfield's ideas have informed algorithms that help reconstruct images from incomplete or noisy scans, improving diagnostic accuracy.

The breakthroughs by Hinton and Hopfield catalyzed a series of innovations that expanded the scope of AI research. Today, AI systems trained with deep learning techniques are capable of performing at—or exceeding—human levels in tasks such as speech recognition, game playing, and visual perception. These advancements are directly traceable to the contributions made by Hinton and Hopfield, whose work transformed AI from an esoteric academic discipline into a technology with practical applications in nearly every industry.

Personal Journeys of the Laureates

The journey of both laureates was far from easy. Geoffrey Hinton's fascination with neural networks began early in his academic career, despite a widespread belief in the AI community that neural networks were a dead end. Hinton's dogged persistence in pursuing neural networks, even when funding and support were sparse, exemplifies his vision for the future of machine learning. It was Hinton's unwavering belief that machines could mimic the brain's learning processes that led him to explore ideas that others had abandoned.

Similarly, John Hopfield's contributions to AI were not immediately embraced. His work, straddling the line between physics, neuroscience, and computation, was initially seen as too unconventional by many in the established fields. Hopfield's interdisciplinary approach to problem-solving, however, allowed him to cross boundaries that others could not, and his neural networks demonstrated how physics could offer valuable insights into cognitive science.

Both Hinton and Hopfield shared a commitment to advancing neural networks despite numerous challenges. Their work not only revolutionized AI but also inspired generations of researchers who continue to push the boundaries of what machines can achieve.

Inspiring the Leading Lights of AI

The role that Geoffrey Hinton in particular has played in putting AI on the map is undeniable. In his long academic career, Hinton has mentored several prominent figures in artificial intelligence who have gone on to make significant contributions to the field. Some of his most famous students include:

1. **Yann LeCun**: Currently the Chief AI Scientist at Meta (formerly Facebook), Yann LeCun is renowned for his work in convolutional neural networks (CNNs) and his pioneering contributions to computer vision and deep learning. He is one of the key figures behind the resurgence of neural networks and deep learning.

2. **Ilya Sutskever**: Co-founder and Chief Scientist at OpenAI, Sutskever is one of the leading researchers in AI, contributing significantly to the development of generative models like GPT-3 and other transformer-based

architectures. His work on sequence learning and optimization is highly influential in deep learning.

3. **Alex Krizhevsky**: Famous for creating AlexNet, the deep convolutional neural network that won the ImageNet competition in 2012, Alex Krizhevsky's work was a breakthrough moment for deep learning in image recognition. He was a PhD student under Geoffrey Hinton at the University of Toronto.

4. **Demis Hassabis**: Co-founder and CEO of DeepMind, Demis Hassabis has been instrumental in the development of AlphaGo, AlphaZero, and other AI systems that have achieved remarkable success in games and scientific challenges like protein folding. Although not a direct student of Hinton, Hassabis has worked closely with him, particularly after DeepMind acquired Hinton's AI company, DNNresearch.

5. **Andrej Karpathy**: Former Director of AI at Tesla and currently a research scientist at OpenAI, Karpathy has been influential in applying deep learning to autonomous driving and AI research. He was heavily inspired by Hinton's work and has worked on recurrent neural networks and computer vision.

The remarkable success of Geoffrey Hinton's students, such as Yann LeCun, Ilya Sutskever, and Alex Krizhevsky, highlights the transformative power of effective mentoring in the field of artificial intelligence. Hinton's mentorship has not only shaped the careers of these individuals but also fundamentally influenced the direction of AI research. By encouraging independent thought, guiding his students through complex research, and fostering an environment of collaboration, Hinton enabled his mentees to make groundbreaking contributions to AI.

Longer-term Impact

The contributions of Geoffrey Hinton and John Hopfield are shaping the future in ways that would have been unimaginable just a few decades ago. Today, deep learning is used to power self-driving cars, translate languages in real-time, recommend products and media, and even assist in medical diagnoses. The legacy of their work is not limited to any one field—its influence spans multiple industries, including healthcare, finance, transportation, and entertainment.

AI systems developed using neural networks are now an integral part of personalized medicine, where algorithms analyze genetic data to predict individual responses to treatment. In finance, AI models optimize trading strategies and manage risk. The work of Hinton and Hopfield has even begun to influence legal systems, where machine learning models are used to predict case outcomes and assist with legal research.

As AI technology continues to evolve, the foundational contributions of Hinton and Hopfield will remain essential. Their breakthroughs have not only transformed industries but have also opened new avenues of research in fields such as quantum computing, ethics in AI, and human-computer interaction.

Fun Facts About Previous Laureates in Physics

The Nobel Prize in Physics has a rich history of recognizing transformative ideas. Some notable previous winners include:

- **Albert Einstein (1921):** Best known for his theory of relativity, Einstein's work redefined our understanding of space, time, and gravity.
- **Marie Curie (1903):** The first woman to win a Nobel Prize, Curie's groundbreaking work on radioactivity laid the foundation for the modern field of nuclear physics.

- **Richard Feynman (1965):** Feynman's contributions to quantum electrodynamics changed the way we understand the interaction of light and matter. He was also celebrated for his engaging teaching style and ability to make complex topics accessible

Check out our Great Scientists series of biographies on each of these notable winners! More details at the end of this book.

Chapter 4: Chemistry

Nobel Prize in Chemistry 2024

Category: Chemistry

Winners: David Baker, Demis Hassabis, John M. Jumper

Associated Institutions: University of Washington (Baker), Google DeepMind (Hassabis and Jumper)

Summary: Baker, Hassabis, and Jumper were awarded the 2024 Nobel Prize in Chemistry for their work on predicting and designing protein structures. Baker was recognized for his achievements in creating entirely new types of proteins. Hassabis and Jumper developed an AI model that solved a 50-year-old problem of predicting complex protein structures. Their discoveries have significant potential for various applications in science and medicine

Detailed Explanation

The 2024 Nobel Prize in Chemistry was awarded to David Baker, Demis Hassabis, and John M. Jumper for their pioneering work in computational protein design and structure prediction. Their achievements have revolutionized the field of protein science, opening new possibilities for understanding biological systems, drug discovery, and personalized medicine. The work of these laureates stands at the intersection of chemistry, biology, and artificial intelligence (AI), demonstrating the profound impact that interdisciplinary approaches can have on solving long-standing scientific challenges.

Transforming Protein Science with AI

When the Nobel Prize committee announced the 2024 laureates, it marked a groundbreaking moment where AI, typically associated with technology and data science, was being recognized for its transformative role in chemistry and biology. The work of Baker, Hassabis, and Jumper exemplifies how AI, specifically deep learning, can be applied to biological sciences to address one of the most complex problems in the field—predicting protein structures.

Proteins are fundamental to life. They serve as enzymes, structural components, and signaling molecules, and their three-dimensional (3D) structure dictates their function. However, understanding this structure has been one of the most significant challenges in biology for decades. The laureates' work has fundamentally changed the way scientists approach protein folding and design, providing faster, more accurate methods to predict and create protein structures.

The Foundations of Protein Research

Proteins are made up of long chains of amino acids that fold into specific 3D shapes, and these shapes determine their function. Predicting how a protein will fold based on its amino acid sequence—known as the protein folding problem—has been a major scientific challenge since the mid-20th century. Historically, determining protein structures required laborious experimental techniques such as X-ray crystallography, nuclear magnetic resonance (NMR) spectroscopy, and cryo-electron microscopy. While effective, these methods were time-consuming and expensive, limiting the speed at which new structures could be understood.

David Baker's work in computational protein design began to change this landscape in the 1990s. He developed the **Rosetta** software, which uses a combination of known protein structures and fragment-based modeling to design new proteins with specific shapes and functions. Baker's approach allowed scientists to create novel proteins that catalyze reactions or carry out functions not found in nature. His research laid the groundwork for the broader application of computational methods in protein science, particularly in drug design and synthetic biology.

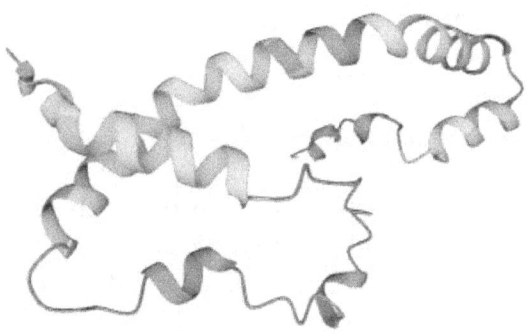

Fig. The Intricate Folds of a Protein
(Source: Wikimedia Commons)

The real breakthrough in protein structure prediction, however, came with the development of **AlphaFold** by Demis Hassabis and John M. Jumper at **Google DeepMind**. AlphaFold, an AI-powered tool, uses deep learning algorithms to predict protein structures with remarkable accuracy based solely on the amino acid sequence. This advancement solved a problem that had confounded scientists for over 50 years, dramatically speeding up the process of understanding protein folding and enabling new research opportunities across biology and chemistry.

David Baker: A Pioneer in Computational Protein Design

David Baker's contributions to protein science extend beyond his work on Rosetta. His laboratory at the University of Washington has been at the forefront of designing proteins from scratch—an endeavor known as *de novo* protein design. Unlike traditional protein engineering, which modifies existing proteins, *de novo* design creates entirely new proteins with desired functions.

For example, Baker's team successfully designed proteins that catalyze chemical reactions not observed in nature. These engineered enzymes have the potential to be used in a wide range of applications, from green chemistry to pharmaceuticals. One of the most notable aspects of Baker's research is its practical application in medicine, particularly in the design of proteins that can neutralize viruses, which holds promise for vaccine development.

The potential of computational protein design is vast. By precisely engineering proteins, scientists can create materials with unique properties, design new therapies for diseases, and even construct nanoscale machines for industrial applications. Baker's work has expanded the horizons of protein science, offering a glimpse into a future where protein-based solutions address some of the most pressing challenges in health and the environment.

Demis Hassabis, John M. Jumper, and the AlphaFold Revolution

Demis Hassabis and John M. Jumper's work on **AlphaFold** represents one of the most significant scientific breakthroughs of the 21st century. Prior to AlphaFold, predicting the 3D structure of a protein from its amino acid sequence was a slow and often imprecise process. Experimental methods were the gold standard, but they required substantial time and resources. The introduction of AlphaFold changed this by making structure prediction faster, more accessible, and more accurate.

AlphaFold's neural network model was trained on vast datasets of known protein structures, allowing it to learn the underlying patterns that dictate how amino acid sequences fold into 3D structures. The result was an AI system that could predict protein structures with astonishing accuracy, often matching or exceeding the performance of traditional experimental methods. AlphaFold's predictions have been validated against experimental data, and its models are now widely used in scientific research.

One of AlphaFold's most significant contributions is its **protein structure database**, which contains predictions for nearly all known proteins. This resource has become invaluable for researchers in fields ranging from drug development to evolutionary biology. By providing detailed models of proteins that had never been experimentally characterized, AlphaFold has enabled scientists to explore new areas of biology and chemistry that were previously inaccessible.

From Experimental to Computational Approaches

The breakthroughs achieved by Baker, Hassabis, and Jumper are the culmination of decades of work in protein science. In the early 20th century, biologists began to unravel the mysteries of protein function and structure, leading to the development of methods like X-ray crystallography in the 1950s. These techniques were critical in determining the structures of proteins such as hemoglobin and myoglobin, but they required years of effort for each protein.

The rise of computational biology in the late 20th century introduced new possibilities for protein research. The development of tools like Rosetta and AlphaFold demonstrated the power of computational methods to complement experimental techniques, offering faster and more scalable ways to study proteins. What was once a slow, resource-intensive process has now become a more efficient and widely accessible approach, thanks to AI-driven models.

Real-World Applications and Impact

The contributions of Baker, Hassabis, and Jumper have far-reaching implications across various fields:

- **Drug Discovery**: Understanding protein structures is essential for developing new drugs, as many pharmaceuticals work by interacting with proteins in the body. AlphaFold's ability to predict structures enables researchers to identify potential drug targets more efficiently.
- **Personalized Medicine**: With the ability to predict individual protein structures, personalized treatments based on a person's unique protein profile are now more feasible, allowing for more targeted therapies.
- **Biotechnology**: In industrial applications, engineered proteins can be used to perform novel functions, such as breaking down plastics or converting carbon dioxide into useful materials. These advancements could lead to more sustainable industrial processes.

Personal Journeys of the Laureates

David Baker, Demis Hassabis, and John M. Jumper come from diverse scientific backgrounds but share a common goal of solving complex biological problems. Baker's journey began with a passion for understanding the fundamental mechanics of proteins. Over the years, his laboratory became a hub for computational biology, developing tools that are now used by scientists worldwide.

Hassabis and Jumper, on the other hand, brought their expertise in AI and machine learning to tackle one of the most challenging problems in biology. Their work at Google DeepMind demonstrated the potential of AI not just in technology but also in scientific research, bridging the gap between computer science and biology.

Inspiring Future Generations

The work of Baker, Hassabis, and Jumper is a testament to the power of collaboration between disciplines. Their breakthroughs demonstrate that the intersection of AI and biology is a fertile ground for innovation, inspiring future generations of scientists to explore new ways of solving age-old problems. As AI technology continues to evolve, the foundational work of these laureates will remain a cornerstone in the ongoing effort to understand and manipulate biological systems for the betterment of humanity.

Long-Term Impact

The research of the 2024 Nobel laureates is already reshaping industries, from healthcare to biotechnology. Their innovations are setting the stage for future advancements in disease treatment, sustainable industrial practices, and our fundamental understanding of life at the molecular level. As AI becomes more integrated into scientific research, the legacy of Baker, Hassabis, and Jumper will continue to influence the trajectory of biology and chemistry for decades to come.

Fun Facts About Previous Laureates in Chemistry

Here are some amazing details about previous Chemistry laureates:

- **Ahmed Zewail (1999)**: Known as the "father of femtochemistry," Zewail won the Nobel Prize for his pioneering work on chemical reactions at the atomic level using extremely fast laser flashes—measured in femtoseconds (one quadrillionth of a second). His work allowed scientists to observe chemical reactions in real-time, earning him a unique spot in history for "filming" molecules in action.

- **Frédéric Joliot-Curie and Irène Joliot-Curie (1935)**: This husband-and-wife team (Irène being the daughter of Marie Curie, whom we mentioned in the end of the previous Chapter) won the Nobel Prize for their discovery of artificial radioactivity. They managed to synthesize new radioactive elements in the lab, a groundbreaking achievement that paved the way for advances in nuclear physics and medicine.

- **Linus Pauling (1954)**: Pauling is the only person to win two unshared Nobel Prizes—one in Chemistry for his research on the nature of the chemical bond and another in Peace for his activism against nuclear weapons testing. His Chemistry prize was awarded for discovering how atoms bond together, which has had a lasting influence on modern chemistry and molecular biology.

- **Dorothy Crowfoot Hodgkin (1964)**: Hodgkin was awarded the Nobel Prize in Chemistry for her work on X-ray crystallography, which she used to determine the structures of important biochemical substances like penicillin, vitamin B12, and later, insulin. Despite suffering from severe rheumatoid arthritis for most of her career, she made groundbreaking contributions that revolutionized medicine and biology. She remains one of the few women to win the Nobel Prize in Chemistry.

Chapter 5: Literature

Nobel Prize in Literature 2023

Category: Literature

Winner: Han Kang

Associated Institution: N/A

Summary: Han Kang was awarded the 2024 Nobel Prize in Literature for her intense poetic prose that confronts historical traumas and exposes the fragility of human life. She is the first South Korean author to receive this prestigious award. Han's works, such as "The Vegetarian" and "Human Acts," are known for their tender yet brutal exploration of themes like trauma, pain, and loss, as well as the connection between the physical and spiritual, the living and the deceased.

Detailed Explanation

The 2024 Nobel Prize in Literature was awarded to **Han Kang**, the renowned South Korean author whose body of work probes deeply into themes of human violence, trauma, identity, and the intersection of the individual and collective psyche. Han's writing transcends national and cultural boundaries, touching on universal aspects of the human condition while rooted in the specific historical and social contexts of Korea. Her works are characterized by a unique blend of poetic sensitivity and stark realism, offering a raw, visceral portrayal of human existence. Her exploration of existential questions and societal norms has earned her international acclaim, making her a voice of global literature. This chapter will explore Han Kang's literary contributions, the environment that shaped her writing, and the far-reaching influence her work continues to have in the literary world.

When the Nobel announcement was made, Han Kang's selection was celebrated widely, with many acknowledging how it represented a broader recognition of Asian literature, especially voices that are less represented in global literary circles. Han's work explores themes that, while deeply personal and culturally specific, resonate with readers around the world. She tackles complex issues like the search for meaning, survival amidst violence, and the limitations of language in expressing profound human emotions. These universal themes are woven into the tapestry of Korea's turbulent history and its societal pressures, making her works a significant reflection on both the personal and the political.

The Foundations of Han Kang's Literary Works

At the core of Han Kang's literature lies a deep interrogation of the fragility of life and the inner workings of the human soul. Her writing is marked by its raw emotional depth, often focusing on marginalized or silenced individuals whose stories reflect broader societal struggles. Han's novels are richly layered, blending stark realism with lyrical prose, allowing readers to navigate the complexities of identity, memory, and trauma. Her ability to balance the visceral with the poetic makes her work both unsettling and beautiful, a signature style that has captivated readers worldwide.

The Vegetarian, one of her most iconic and celebrated works, tells the story of a woman who chooses to reject meat in a society that values conformity and tradition. What begins as a personal decision spirals into a much larger exploration of bodily autonomy, mental instability, and societal repression. The protagonist's quiet rebellion sets off a chain reaction, leading to the disintegration of her relationships and her descent into a psychological breakdown. The novel is a profound meditation on desire, autonomy, and the body, and it garnered international recognition for its exploration of these complex issues.

Another one of Han Kang's critically acclaimed works, **Human Acts**, deals with the 1980 Gwangju Uprising, a crucial event in South Korea's modern history. Through multiple perspectives, Han traces the physical and emotional toll of this brutal crackdown on pro-democracy protesters. The novel's fragmented narrative structure echoes the fractured memories of those affected by the violence, offering a poignant reflection on how trauma shapes both individuals and nations. **Human Acts** not only confronts the horror of political violence but also speaks to the lingering pain of historical memory, making it one of her most powerful works.

Han Kang's Exploration of Trauma and Identity

A central theme in Han Kang's writing is trauma—both personal and collective—and its profound impact on identity. She closely examines the emotional and psychological scars left by violence, examining how individuals process pain, loss, and alienation. Her novels often feature protagonists who experience disconnection from their bodies, their environments, or their identities, reflecting the ways in which trauma fragments the self. In **Human Acts**, the victims of the Gwangju Uprising struggle not only with physical wounds but with the intangible scars of memory and grief. The novel's shifting perspectives illustrate how trauma reverberates through time, affecting not only those who directly experienced the violence but also future generations.

In **The White Book**, Han adopts a more introspective and poetic approach to the exploration of trauma. The novel reflects on her personal grief over the death of her sister, who died shortly after birth, through a meditation on the color white. Each chapter presents fragmented images—snow, breast milk, fog—that serve as metaphors for life, death, and memory. The novel is less a narrative and more an elegy, offering readers a haunting exploration of how loss leaves its mark on the living. In this way, **The White Book** serves as a powerful testament to Han's ability to make the personal universal, transforming individual grief into a meditation on the human experience of mourning and remembrance.

Historical and Social Context

Han Kang's writing is deeply shaped by South Korea's tumultuous political history and rapid modernization. Born in the city of Gwangju in 1970, Han grew up in the shadow of the Gwangju Uprising, a pro-democracy movement that was violently suppressed by the military dictatorship in 1980. This event is central to **Human Acts**, which grapples with the brutality of the regime and the long-lasting emotional scars left on the survivors and their families. South Korea's transition from authoritarianism to democracy, along with the cultural shifts that accompanied the country's economic boom, provides a rich backdrop for Han's exploration of trauma, memory, and identity.

The themes of silence and repression, both individual and collective, are prevalent in her work, often reflecting the societal pressures that define South Korean life. Han uses her novels as a means of exploring the consequences of social conformity and the repression of individual desire. In **The Vegetarian**, the protagonist's refusal to conform to societal norms by rejecting meat serves as a metaphor for personal rebellion against the cultural expectations placed on women. Her gradual mental unraveling can be seen as a commentary on the psychological toll of living in a society that stifles individuality and enforces strict social roles.

Han Kang's work also speaks to broader global themes of violence, survival, and the human condition. While deeply rooted in the specific context of South Korea's political history and societal norms, her exploration of trauma and identity has resonated with readers around the world. Her works confront readers with the uncomfortable realities of violence and repression, yet they also offer moments of profound beauty and redemption.

Personal Journey

Han Kang was born into a literary family; her father, Han Seung-won, is a well-known novelist in South Korea. Growing up in a literary household, Han was immersed in the world of storytelling from an early age. She studied Korean literature at Yonsei University, one of South Korea's most prestigious institutions, and began her career as a poet before transitioning to fiction.

Her early works were well-received in South Korea, but it was with **The Vegetarian** that she gained international recognition. After winning the **Man Booker International Prize** in 2016, Han's reputation as one of the leading voices in world literature was cemented. Despite her global success, Han has remained committed to writing deeply personal and introspective works, often shying away from the public eye.

Han's journey as a writer has been marked by her dedication to exploring the darker aspects of the human psyche. Her protagonists are often outsiders, individuals who struggle to fit into the rigid frameworks imposed by society. Whether it's a woman rejecting societal expectations in **The Vegetarian**, or victims of political violence in **Human Acts**, Han's characters are defined by their sense of isolation and their search for meaning in a world that often seems indifferent to their suffering.

Impact and Legacy

Han Kang's work has left an indelible mark on the literary world. Her novels have been translated into multiple languages, making her one of the most widely read South Korean authors internationally. By addressing themes of trauma, violence, and the human condition, Han has brought South Korean literature to a global audience, offering readers a glimpse into the complexities of Korean history and culture while engaging with universal themes.

Her success has also paved the way for other South Korean writers to gain international recognition. In recent years, there has been a surge of interest in South Korean literature, with authors like **Kim Young-ha** and **Bae Suah** following in Han's footsteps. Han's ability to tackle difficult, often uncomfortable subjects has opened new avenues for storytelling, encouraging writers to confront issues of trauma, identity, and societal repression head-on.

Inspiring a New Generation

Han Kang's literary success and bold thematic choices have inspired a new generation of writers both in South Korea and internationally. Her exploration of trauma, memory, and societal norms encourages younger authors to push the boundaries of conventional storytelling. Han's influence can be seen in the works of contemporary South Korean writers who are similarly engaged with themes of identity, history, and social expectation.

In addition to inspiring writers, Han has contributed to a broader cultural conversation about the role of literature in addressing societal and historical trauma. Her novels serve as a reminder of the power of storytelling to give voice to the silenced, challenge societal norms, and promote empathy among readers. Han Kang's literary legacy is not only one of artistic achievement but also one of social and cultural relevance.

Conclusion

The 2024 Nobel Prize in Literature recognizes **Han Kang** for her profound contributions to global literature. Her exploration of trauma, memory, and human resilience has not only enriched South Korean literature but has also made an indelible impact on the world stage. Through her deeply introspective and emotionally charged narratives, Han Kang offers readers a window into the complexities of the human spirit, making her a deserving laureate for one of the highest honors in literature.

Fun Facts About Previous Laureates in Literature

Here are some fascinating details about past Nobel Literature Prize winners:

- **Gabriel García Márquez (1982)**: Often referred to as the master of magical realism, García Márquez's *One Hundred Years of Solitude* is considered one of the most influential novels of the 20th century. Interestingly, when he was awarded the Nobel Prize, García Márquez flew to Stockholm with 300 guests from his hometown and insisted on wearing a traditional Colombian *liquiliqui* instead of a tuxedo at the ceremony.
- **Bob Dylan (2016)**: As a rare laureate not from traditional literature, Bob Dylan was awarded the Nobel Prize for creating "new poetic expressions within the great American song tradition." Initially, Dylan did not respond to the Nobel committee's calls and emails, creating a brief controversy. He also skipped the ceremony, sending a speech to be read in his absence.
- **Rabindranath Tagore (1913)**: Tagore was the first non-European to win the Nobel Prize in Literature. His win put Indian literature in the international spotlight. Fun fact: He was also a musician and composed both the Indian national anthem (*Jana Gana Mana*) and the Bangladeshi national

anthem (*Amar Shonar Bangla*), making him the only laureate to have penned national anthems for two countries.
- **Doris Lessing (2007)**: When the Swedish Academy announced that she had won the Nobel Prize, Doris Lessing was famously caught by surprise while getting out of a taxi. Her reaction, "Oh Christ... I couldn't care less," became a widely shared moment. Known for her works like *The Golden Notebook*, Lessing was awarded the prize at the age of 88, making her the oldest recipient of the Nobel Prize in Literature at the time.

These laureates not only shaped the literary world with their extraordinary works, but their personal stories and unique paths to success offer insight into the diverse range of voices that the Nobel Committee has honored over the years.

Chapter 6: Peace

Nobel Peace Prize 2024

Category: Peace

Winner: Nihon Hidankyo

Associated Institution: Nihon Hidankyo (Japan Confederation of A- and H-Bomb Sufferers Organizations

Summary: Nihon Hidankyo, a grassroots movement of atomic bomb survivors from Hiroshima and Nagasaki (also known as Hibakusha), was awarded the 2024 Nobel Peace Prize for its efforts to achieve a world free of nuclear weapons and for demonstrating through witness testimony that nuclear weapons must never be used again. The award was given amidst concerns that the taboo against the use of nuclear weapons is under pressure.

Detailed Explanation

The 2024 Nobel Peace Prize was awarded to Nihon Hidankyo, a Japanese organization representing survivors of atomic bombings, known as hibakusha. The group was recognized for its relentless advocacy towards the abolition of nuclear weapons, promoting peace, and preserving the memory of the devastating impact of nuclear war. Nihon Hidankyo's dedication has kept the dialogue on nuclear disarmament alive for decades, offering firsthand testimony of the horrors that nuclear weapons bring. This Nobel Prize marks a significant acknowledgment of the human cost of nuclear warfare and serves as a global call to action for disarmament. It is also a clarion call to future generations to be aware of the ever-present nuclear danger.

A Voice for the Survivors

The Nobel Peace Prize committee's choice to honor Nihon Hidankyo symbolizes an acknowledgment of the hibakusha's suffering and resilience. Nihon Hidankyo, formed in 1956, has worked tirelessly to bring the world's attention to the consequences of nuclear bombings. They have shared their personal experiences to highlight the human and environmental costs of nuclear warfare, giving a human face to what often remains an abstract issue for many. Their advocacy has been instrumental in shaping the global discourse on nuclear weapons, influencing policymakers, international bodies, and civil society alike.

Fig. The iconic logo of Nihon Hidankyo

The hibakusha—the survivors of Hiroshima and Nagasaki—have consistently shared their tragic experiences to emphasize the catastrophic effects of nuclear weapons. Through public speaking engagements, documentaries, and international conferences, Nihon Hidankyo has given these survivors a platform to speak. The organization's efforts have helped foster international solidarity against nuclear arms, inspiring numerous initiatives aimed at achieving a world without nuclear weapons.

Advocating for Nuclear Disarmament

Nihon Hidankyo's advocacy work has been grounded in the principles of peace, human dignity, and nuclear disarmament. They have been vocal proponents of the Treaty on the Prohibition of Nuclear Weapons (TPNW), which was adopted by the United Nations in 2017. This treaty aims to legally prohibit nuclear weapons, leading towards their complete elimination. The hibakusha played a critical role in the treaty's adoption by sharing their stories, which humanized the urgency of nuclear disarmament for those who might otherwise view it as a geopolitical or technical issue.

The hibakusha have consistently emphasized that their goal is not only to prevent the use of nuclear weapons again but also to eliminate their very existence. Nihon Hidankyo has been at the forefront of pressing nations—especially nuclear-armed ones—to engage in disarmament talks and commit to reducing their nuclear arsenals. Their advocacy emphasizes not just the prevention of nuclear war but also the inherent dangers in maintaining a global nuclear arsenal.

Survivors as Symbols of Resilience

The members of Nihon Hidankyo have not only advocated for disarmament but also provided hope for a peaceful future through their resilience. Many hibakusha have suffered lifelong health issues due to radiation exposure, and yet, they have committed their lives to ensuring that the horrors of Hiroshima and Nagasaki are never repeated. Their endurance and commitment have inspired individuals and organizations worldwide to support disarmament efforts, emphasizing the importance of empathy in political discourse.

The hibakusha's testimony serves as a stark reminder of the destructive power of nuclear weapons. Their message is simple yet profoundly impactful: no one else should endure the suffering that they have experienced. By continuing to share their experiences despite their advancing age, these survivors have provided an irreplaceable voice in the fight for global disarmament and peace.

Building Bridges for Global Peace

Nihon Hidankyo has also worked to build alliances with peace organizations and anti-nuclear advocates around the world. The hibakusha have collaborated with international bodies, including the United Nations and non-governmental organizations (NGOs), to amplify their message and create a global network pushing for nuclear disarmament. Their partnership with like-minded groups has fostered a greater understanding of the necessity of nuclear disarmament, not just among governments but also among ordinary citizens worldwide.

One of Nihon Hidankyo's major contributions is its role in education—particularly for younger generations. By emphasizing the importance of historical memory, they have inspired students, educators, and activists to engage in peace advocacy. This educational outreach ensures that the tragic lessons of Hiroshima and Nagasaki are not forgotten, and it builds a foundation for a future without nuclear weapons.

The Path to Disarmament

The recognition of Nihon Hidankyo with the 2024 Nobel Peace Prize comes at a time when global tensions around nuclear arms are on the rise. The prize highlights the urgency of addressing nuclear disarmament as a critical issue of our time. Nihon

Hidankyo's persistence shows that the moral argument against nuclear weapons remains powerful, despite the geopolitical complexities surrounding disarmament efforts. Their work serves as a poignant reminder that nuclear weapons are not just tools of deterrence but instruments of unimaginable human suffering.

This award not only celebrates Nihon Hidankyo's tireless advocacy but also reinvigorates the global movement for a nuclear-free world. The Nobel Committee's decision is an affirmation of the hibakusha's message: that the memory of past horrors must serve as the foundation for a peaceful and nuclear-free future.

Fun Facts About Previous Laureates in Peace

Here are some interesting details about previous Nobel Peace Prize laureates:

- **Mother Teresa (1979)**: Mother Teresa was awarded the Nobel Peace Prize for her humanitarian work with the poor and dying. Rather than attending the traditional banquet, she requested that the $192,000 fund be used to help the needy in India.
- **Malala Yousafzai (2014)**: At 17 years old, Malala became the youngest-ever recipient of the Nobel Peace Prize. She was recognized for her struggle against the suppression of children and young people and for the right of all children to education, following her courageous activism after surviving an assassination attempt by the Taliban.
- **Albert Schweitzer (1952)**: Schweitzer received the Nobel Peace Prize for his philosophy of "Reverence for Life," which he used as the basis for his humanitarian mission in Africa, where he established a hospital in Gabon. He was both a physician and a musician, known for his work as an organist and his studies of J.S. Bach.
- **Nelson Mandela and Frederik Willem de Klerk (1993)**: Mandela and de Klerk shared the Nobel Peace Prize for

their efforts to end apartheid in South Africa. Their partnership was crucial in dismantling the apartheid regime and laying the foundation for a democratic South Africa, highlighting the power of reconciliation and cooperation.

Chapter 7: Economics

Nobel Peace Prize in Economic Sciences 2024

Category: Economics

Winners: Daron Acemoglu, Simon Johnson, and James A. Robinson

Associated Institutions: Massachusetts Institute of Technology (Acemoglu, Johnson), University of Chicago (Robinson)

Summary: Acemoglu, Johnson, and Robinson were awarded the 2024 Nobel Prize in Economics for their studies on how institutions are formed and affect prosperity. The three economists have demonstrated the importance of societal institutions for a country's prosperity through their influential collaborations on the relationship between political institutions, economic development, and long-term prosperity.

Detailed Explanation

The 2024 Nobel Prize in Economics was awarded to Daron Acemoglu, Simon Johnson, and James A. Robinson for their studies of how institutions are formed and affect prosperity. Their pioneering research has demonstrated the importance of societal institutions for a country's prosperity, providing insights into the role of political and economic systems in fostering or hindering growth. Acemoglu, Johnson, and Robinson's work has significantly influenced our understanding of the historical and institutional factors that determine the economic outcomes of nations. This chapter will explore their discoveries, the historical context that preceded their breakthroughs, and the profound impact their work continues to have on economic thought and policy-making.

When the announcement was made, many observers lauded the Nobel Committee's choice for recognizing economists who have helped us understand differences in prosperity between nations. In an era where global inequality remains a pressing issue, the laureates' work stood out for its ability to explain why some countries thrive while others struggle to achieve sustained economic growth.

Acemoglu, Johnson, and Robinson have leveraged historical data, case studies, and advanced econometric techniques to analyze the role of institutions in economic development. Their research illustrates how institutions that promote inclusive political and economic systems foster long-term prosperity, while extractive institutions hinder growth and maintain inequality. The committee specifically recognized their contributions for highlighting the significance of institutional quality in shaping economic outcomes and development trajectories.

The Foundations of Institutional Economics

Institutional economics seeks to understand how different types of institutions—such as legal, political, and economic structures—affect economic performance and development. While traditional economic theory often focused on market mechanisms, the work of Acemoglu, Johnson, and Robinson exemplifies a shift towards understanding the broader socio-political context that influences economic outcomes. Their research provides a framework for analyzing how institutional arrangements can either promote or hinder development, depending on their inclusivity or extractiveness.

Why Institutions Matter?

Daron Acemoglu, a professor at MIT, has been a leading figure in institutional economics, focusing on how political and economic institutions shape economic outcomes. Acemoglu's work, particularly in collaboration with Robinson, has provided a theoretical and empirical basis for understanding why some nations succeed while others remain trapped in poverty. Their influential book "Why Nations Fail" argues that inclusive institutions—those that provide broad access to resources and opportunities—are essential for sustained economic growth.

One of Acemoglu's notable contributions is his analysis of the impact of historical events, such as colonization, on the development of institutions. His research shows that the institutions established during colonial times have had lasting effects on the prosperity of former colonies. For instance, regions where colonizers established inclusive institutions, with property rights and political representation, are today more prosperous compared to those where extractive institutions were designed to exploit the local population.

Acemoglu's dedication to understanding the role of institutions has transformed how economists think about development. By emphasizing the importance of inclusive governance and equitable economic systems, his work has inspired numerous researchers and policymakers to focus on institutional reforms as a means to foster economic growth.

Effects of Institutional Change

Simon Johnson, also a professor at MIT, has focused on the effects of institutional change on economic development, particularly in the context of emerging markets. Johnson's research has examined how shifts in political power and institutional arrangements affect economic stability and growth. One of his key contributions is his work on how weak institutions can lead to financial crises and hinder recovery.

Johnson's work on the economic history of colonization, particularly in collaboration with Acemoglu, has highlighted the role of extractive institutions in perpetuating poverty and inequality. By analyzing historical data, Johnson has shown that countries with a legacy of extractive institutions are more likely to experience poor economic performance, corruption, and instability. His research underscores the need for institutional reforms that promote transparency, accountability, and broad-based economic participation.

Through his work, Johnson has contributed to a better understanding of how institutional quality affects economic resilience. His insights have been particularly influential in shaping policy debates around governance, economic reforms, and the role of international organizations in supporting institutional development in low-income countries.

The Role of Political Economy

James A. Robinson, a professor at the University of Chicago, has extensively studied the interplay between political power and economic institutions. Robinson's research has focused on how political dynamics shape economic policies and how these, in turn, affect development outcomes. His collaboration with Acemoglu has led to influential theories on why some nations develop inclusive institutions while others become trapped in cycles of extractive governance.

Robinson's work has highlighted the importance of political stability and the distribution of power in shaping institutions. He has argued that inclusive political systems, which distribute power broadly and limit the ability of elites to exploit the system, are key to achieving long-term economic prosperity. Conversely, extractive political systems, which concentrate power in the hands of a few, tend to create economic systems that benefit elites at the expense of broader society.

Robinson's contributions have provided a deeper understanding of the role of political incentives in shaping economic institutions. His work has been instrumental in informing policy discussions on democratization, governance, and the need for political reforms that support inclusive economic growth.

Historical Context

The research of Acemoglu, Johnson, and Robinson builds on a long history of efforts to understand the determinants of economic development. Throughout much of the 20th century, economists focused on factors such as capital accumulation, technological progress, and human capital as the primary drivers of growth. However, these models often failed to explain why some countries with similar resources experienced vastly different outcomes.

The shift towards institutional analysis gained momentum in the late 20th century, as economists began to recognize the importance of governance, legal frameworks, and political stability in shaping economic outcomes. The work of Acemoglu, Johnson, and Robinson has been instrumental in this transformation, providing a comprehensive framework for understanding how historical events, such as colonization, have led to the establishment of institutions that continue to influence economic trajectories today.

Their research highlights how the institutions introduced during colonization had a profound impact on the development paths of different regions. In some areas, inclusive institutions were established, fostering long-term growth and prosperity. In others, extractive institutions were put in place, leading to persistent poverty and underdevelopment. This historical perspective has been crucial in explaining the "reversal of fortune" phenomenon, where regions that were once wealthy became impoverished due to the nature of their colonial institutions.

Understanding Institutions and Prosperity

The research by Acemoglu, Johnson, and Robinson has provided a new lens through which to view economic development. By demonstrating that the quality of institutions is a fundamental determinant of economic success, they have challenged traditional economic models that focused primarily on market forces and resource endowments. Their work shows that without inclusive institutions, economic policies are unlikely to succeed in promoting growth and reducing inequality.

One of their key insights is that extractive institutions, which concentrate power and wealth in the hands of a few, create conditions that hinder innovation, investment, and broad-based economic participation. Conversely, inclusive institutions provide the foundation for secure property rights, political representation, and economic opportunities, all of which are essential for sustained growth.

The breakthroughs made by Acemoglu, Johnson, and Robinson have redefined development economics, placing institutions at the center of the analysis. Their work has demonstrated that economic development is not just about accumulating capital or improving technology—it is also about creating the right institutional environment that allows people to thrive.

Personal Journeys of the Laureates

The journeys of all three laureates have been marked by a deep interest in understanding the root causes of economic inequality.

Daron Acemoglu, born in Istanbul, Türkiye, pursued his PhD at the London School of Economics, where he began exploring the role of institutions in economic development. His passion for understanding why some nations prosper while others remain poor led him to collaborate with Johnson and Robinson, resulting in some of the most influential work in the field of development economics.

Simon Johnson, originally from Sheffield, UK, earned his PhD from MIT and has focused much of his career on understanding how institutional weaknesses contribute to economic crises. His work at the International Monetary Fund and his extensive research on emerging markets have given him a unique perspective on the challenges faced by developing economies.

James A. Robinson, educated at Yale University, has spent decades studying political economy and the interplay between political institutions and economic outcomes. His interest in how power dynamics shape economic development has led him to collaborate closely with Acemoglu, resulting in a series of influential publications that have reshaped our understanding of institutional economics.

Inspiring Future Economists

The work of Acemoglu, Johnson, and Robinson has inspired a new generation of economists who seek to understand the deeper forces that shape economic development. Their emphasis on institutions has led to a growing body of research focused on governance, political stability, and the role of historical events in shaping modern economies. Some of their notable collaborators and mentees include:

1. **Suresh Naidu**: A political economist who has worked with Robinson on understanding the role of political power in economic development, particularly in the context of labor markets and inequality.

2. **James Fearon**: Collaborated with Robinson on research related to political instability and its economic consequences, providing insights into how conflicts and power struggles affect institutional development.
3. **Melissa Dell**: A prominent economist whose research on the long-term impact of colonial institutions has been influenced by the work of Acemoglu, Johnson, and Robinson. Her work has expanded on their theories by examining how historical institutions continue to shape economic outcomes today.

The legacy of Acemoglu, Johnson, and Robinson lies not only in their groundbreaking research but also in their commitment to fostering a deeper understanding of the role of institutions in economic development. Their influence can be seen in the growing emphasis on institutional analysis across economic disciplines, with more researchers aiming to use historical and empirical approaches to tackle pressing development challenges.

Longer-term Impact

The contributions of Daron Acemoglu, Simon Johnson, and James A. Robinson are shaping the future of economics by embedding institutional analysis at the core of development theory. Today, policymakers and international organizations increasingly recognize that improving institutional quality is essential for achieving sustainable growth. Their work has informed a wide range of policy initiatives, from governance reforms to anti-corruption measures, aimed at creating more inclusive societies.

The insights provided by the laureates have influenced initiatives aimed at democratization, improving governance, and promoting economic inclusion. Their research on the importance of inclusive institutions has led to significant policy changes, including efforts to strengthen property rights, reduce corruption, and enhance political representation in developing countries. The lasting impact of their work is evident in how policymakers now approach the challenges of development—not just through economic reforms, but by addressing the underlying institutional barriers to progress.

Fun Facts About Previous Laureates in Economics

The Nobel Prize in Economics has a storied history of recognizing transformative work. Some notable previous winners include:

- **Amartya Sen (1998)**: Known for his work on welfare economics, Sen's research has profoundly influenced our understanding of poverty and inequality. His groundbreaking concept of "capabilities" has redefined how we measure well-being beyond just income. As a young boy, Sen grew up on the campus of Santiniketan, a university founded by the poet Rabindranath Tagore (and a former Nobel Laureate in Literature). He was once caught reading a book in class that wasn't assigned—it turned out to be a text on Marxist economics, sparking a lifelong curiosity about inequality.
- **Milton Friedman (1976)**: A leading figure in the Chicago School of Economics, Friedman's work on monetary theory has shaped modern economic policies. Friedman was famous for his love of debate and had a knack for distilling complex economic concepts into simple analogies—like his famous "helicopter drop" idea, which suggested that governments could drop money from helicopters to increase consumer spending during deflationary periods.
- **Elinor Ostrom (2009)**: The first woman to win the Nobel Prize in Economics, Ostrom's research on the governance of common resources has had a significant impact on environmental economics and policy. Ostrom was known for her love of fieldwork—she would wade through streams and hike into forests to better understand how local communities managed their resources, often preferring to observe a community meeting rather than sit through an academic conference.

DR. LEO LEXICON

That concludes our journey reviewing the Nobel Prize Winners of 2024! Congratulations to all the winners!

Now, stay tuned for a FREE excerpt from one of our bestselling biographies from the Great Scientists Series, **EINSTEIN: The Man, the Myth, the Legend**. As you already know, the great Albert Einstein won the Nobel Prize in Physics in 1921 for his discovery of the photoelectric effect, which was the key to establishing quantum theory. Interestingly, he was not awarded for his theory of relativity, which was still considered controversial at the time! I hope you will consider adding it to your library.

**Meet the most well-known Nobel Laureate.
FREE SAMPLE from this book coming up NEXT!**

FREE SAMPLE:

Chapter 3

General Relativity

Understanding Spacetime

Albert Einstein's general theory of relativity, published in 1915, represents one of the most profound and far-reaching intellectual achievements in the history of science. It fundamentally transformed our understanding of gravity, space, and time, and opened up new avenues of research in cosmology, astrophysics, and fundamental physics that continue to be explored to this day.

The genesis of general relativity can be traced back to Einstein's early work on special relativity, which he developed in 1905. Special relativity dealt with the behavior of space and time in inertial reference frames – that is, frames of reference moving at constant velocity relative to each other. It showed that the traditional Newtonian notions of absolute space and time had to be replaced by a new, more flexible framework in which the measurements of space and time were relative to the observer.

But special relativity had a significant limitation: it did not account for the effects of gravity. Einstein recognized that to create a truly comprehensive theory of space, time, and motion, he would need to extend the principles of relativity to accelerated reference frames and find a way to incorporate gravity into the framework.

This was no easy task. Einstein spent nearly a decade struggling to develop a mathematical formalism that could describe gravity in a way that was consistent with the principles of special relativity. He explored a variety of approaches, from simple thought experiments to complex tensor calculus, trying to find a way to reconcile the apparent contradiction between the constancy of the speed of light and the universality of free fall.

One of the key insights that guided Einstein's work was the principle of equivalence, which stated that the effects of gravity were indistinguishable from the effects of acceleration. This meant that an observer in a freely falling elevator would experience the same physical laws as an observer in deep space, far from any gravitational fields.

Einstein realized that this principle had profound implications for the nature of space and time. If gravity and acceleration were equivalent, then the presence of matter and energy must somehow affect the geometry of space-time itself. This led him to the idea that gravity was not a force in the traditional sense, but rather a manifestation of the curvature of space-time caused by the presence of mass and energy.

To develop this idea into a full-fledged theory, Einstein had to master the tools of differential geometry and tensor calculus, which allowed him to describe the curvature of space-time in mathematical terms. He also had to confront a number of conceptual and technical challenges, from the problem of defining a consistent notion of time in curved space-time to the difficulty of incorporating the principles of conservation of energy and momentum into the theory.

Throughout this process, Einstein was guided by a deep physical intuition and a relentless pursuit of a unified, elegant theory that could explain the fundamental workings of the universe. He was also influenced by the work of other scientists and mathematicians, such as the German mathematician David Hilbert, who was independently developing a similar geometrical approach to gravity.

In November 1915, after years of intense intellectual effort, Einstein finally presented his general theory of relativity to the Prussian Academy of Sciences in Berlin. The theory was a masterpiece of mathematical and physical insight, providing a radically new understanding of gravity as the curvature of space-time caused by the presence of matter and energy.

The implications of general relativity were profound and far-reaching. It made a number of startling predictions, from the bending of light by massive objects to the existence of black holes and the possibility of gravitational waves. It also provided a new framework for understanding the large-scale structure and evolution of the universe, paving the way for the development of modern cosmology.

But perhaps most importantly, general relativity represented a triumph of the human intellect and a testament to the power of pure thought to reveal the hidden workings of nature. It showed that even the most abstract and esoteric mathematical concepts could have deep physical meaning and lead to new insights into the fundamental nature of reality.

The genesis of general relativity was a long and arduous process, requiring years of intense intellectual effort and creative problem-solving. But the result was a theory of unparalleled elegance and explanatory power, one that continues to inspire and guide scientific research to this day. It is a testament to Einstein's genius and his unrelenting pursuit of truth, and a reminder of the incredible potential of the human mind to unlock the secrets of the universe.

Key Concepts Explained

At the heart of Einstein's general theory of relativity is the idea that gravity is not a force in the traditional sense, but rather a manifestation of the curvature of space-time caused by the presence of matter and energy. This is a profound and counterintuitive concept, one that requires a radical rethinking of our fundamental notions of space, time, and causality.

To understand the key concepts of general relativity, it is helpful to start with the idea of space-time itself. In traditional Newtonian physics, space and time were considered separate and absolute entities, with space being a fixed, three-dimensional stage on which objects moved and interacted, and time being a universal, one-dimensional flow that proceeded at the same rate for all observers.

But Einstein's special theory of relativity showed that this picture was incomplete. It demonstrated that space and time were intimately connected, and that the measurements of space and time were relative to the observer's state of motion. This led to the idea of a four-dimensional space-time continuum, in which space and time were united into a single geometric entity.

General relativity takes this idea a step further, showing that the geometry of space-time is not fixed and absolute, but rather is determined by the distribution of matter and energy throughout the universe. In other words, massive objects like stars and planets cause space-time to curve and warp around them, and this curvature is what we experience as gravity.

To visualize this, imagine a stretched rubber sheet representing space-time. If you place a heavy object like a bowling ball on the sheet, it will cause the sheet to deform and curve around it. Now imagine rolling a smaller ball across the sheet. Instead of traveling in a straight line, the ball will follow a curved path, as if it were being "pulled" towards the heavy object. This is analogous to how light and matter move in the presence of a gravitational field.

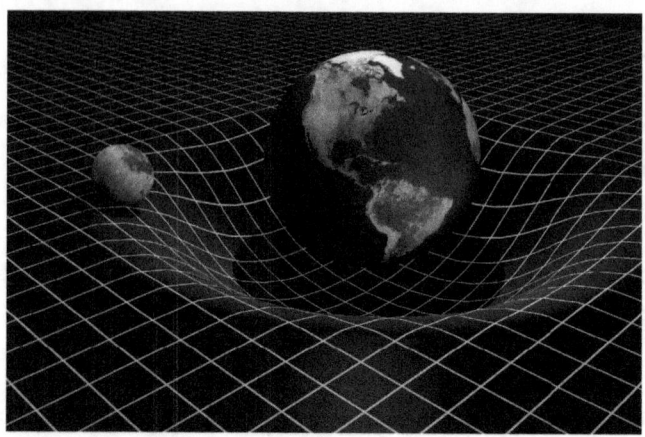

Fig. Visualizing the Curvature of Spacetime

The mathematical description of this curvature is given by the Einstein field equations, which relate the curvature of space-time to the distribution of matter and energy. These equations are notoriously complex and difficult to solve, requiring advanced techniques in differential geometry and tensor calculus.

But the physical implications of the equations are profound. They show that gravity is not a force that acts instantaneously across space, as Newton had imagined, but rather is a consequence of the deformation of space-time itself. This means that gravitational effects can propagate through space-time at the speed of light, in the form of ripples or waves known as gravitational waves.

The existence of gravitational waves was one of the most stunning predictions of general relativity, and one that was not confirmed experimentally until a century after Einstein first proposed it. Many decades after Einstein, and as recently as 2015, the Laser Interferometer Gravitational-Wave Observatory (LIGO) detected the first direct evidence of gravitational waves, produced by the collision of two massive black holes over a billion light-years away.

This detection was a triumph of experimental physics and a powerful confirmation of the predictions of general relativity. It opened up a whole new field of astronomy, allowing scientists to study the universe in a completely new way and to probe some of the most extreme and exotic objects in the cosmos.

Another key concept in general relativity is the idea of geodesics, which are the shortest paths between two points in curved space-time. In flat space-time, geodesics are simply straight lines, but in curved space-time, they can be complex and convoluted, depending on the distribution of matter and energy.

The motion of objects in general relativity is determined by the geodesics of space-time. This means that objects in free fall, like satellites orbiting the Earth or planets orbiting the sun, are actually following the straightest possible paths in curved space-time. It is the curvature of space-time itself that gives rise to the apparent force of gravity, not any intrinsic property of the objects themselves.

This has profound implications for our understanding of the universe on the largest scales. It means that the large-scale structure and evolution of the cosmos is determined by the distribution of matter and energy, and by the resulting curvature of space-time. This has led to the development of modern cosmology, the study of the origin, structure, and ultimate fate of the universe as a whole.

General relativity has also led to the prediction of exotic objects like black holes, which are regions of space-time where the curvature becomes so extreme that not even light can escape. The existence of black holes was long considered a purely theoretical possibility, but in recent years, astronomers have found compelling evidence for their existence, from the detection of gravitational waves to the first direct image of a black hole's event horizon.

These are just a few of the key concepts and implications of general relativity, a theory that has revolutionized our understanding of space, time, and gravity. It is a testament to the power of pure thought and mathematical abstraction to reveal the hidden workings of nature, and a reminder of the incredible depth and richness of the universe we inhabit.

Impact on Physics

The impact of Einstein's general theory of relativity on the field of physics cannot be overstated. It represented a profound and radical departure from the classical Newtonian framework that had dominated scientific thought for centuries, and opened up entirely new avenues of research and discovery that continue to shape our understanding of the universe to this day.

One of the most significant impacts of general relativity was on the study of cosmology, the branch of physics that deals with the origin, structure, and evolution of the universe as a whole. Prior to Einstein's theory, cosmology was a largely speculative and philosophical field, with little in the way of empirical evidence or mathematical rigor.

But general relativity provided a powerful new framework for understanding the large-scale structure of the universe. It showed that the universe was not a static and unchanging entity, as had been previously assumed, but rather was a dynamic and evolving system governed by the laws of gravity and the distribution of matter and energy.

This led to the development of modern cosmological models, such as the Big Bang theory, which posits that the universe began as a singularity of infinite density and temperature, and has been expanding and cooling ever since. The Big Bang theory has been incredibly successful in explaining a wide range of observational data, from the abundance of light elements in the universe to the cosmic microwave background radiation that permeates all of space.

General relativity also had a profound impact on the study of black holes, which are perhaps the most exotic and extreme objects in the universe. Black holes are regions of space-time where the curvature becomes so intense that not even light can escape, and where the laws of physics as we know them break down.

The existence of black holes was first predicted by the German physicist Karl Schwarzschild in 1916, just a few months after Einstein published his theory. But it was not until the 1960s and 1970s that the study of black holes really took off, thanks in large part to the work of physicists like Roger Penrose and Stephen Hawking.

Hawking, in particular, made a number of seminal contributions to the study of black holes, including the discovery of Hawking radiation, which showed that black holes are not completely black, but rather emit a tiny amount of radiation due to quantum effects. This discovery had profound implications for our understanding of the relationship between gravity and quantum mechanics, and helped to spawn the field of quantum gravity.

Another major impact of general relativity was on the study of gravitational waves, which are ripples in the fabric of space-time caused by the acceleration of massive objects. The existence of gravitational waves was first predicted by Einstein himself in 1916, but it was not until a century later that they were finally detected directly, thanks to the incredible precision and sensitivity of modern gravitational wave observatories like LIGO and Virgo.

The detection of gravitational waves has opened up a whole new window on the universe, allowing scientists to study some of the most extreme and violent events in the cosmos, from the collision of black holes and neutron stars to the birth of the universe itself. It has also provided a powerful new tool for testing the predictions of general relativity in the strong-field regime, where the effects of gravity are most intense.

But perhaps the most profound impact of general relativity has been on our fundamental understanding of the nature of space, time, and gravity. It has shown that these concepts are not fixed and absolute, but rather are dynamical and interconnected, shaped by the presence of matter and energy.

This has led to a profound shift in our philosophical and metaphysical understanding of the universe, and has challenged many of our most deeply held beliefs about the nature of reality. It has also inspired countless new theories and models, from string theory and loop quantum gravity to the holographic principle and the multiverse.

In short, general relativity has been one of the most influential and transformative theories in the history of physics. It has shaped our understanding of the universe on the largest scales, and has opened up entirely new fields of research and discovery. It is a testament to the power of human curiosity and creativity, and a reminder of the incredible beauty and complexity of the cosmos we inhabit.

Controversies and Acceptance

Despite its incredible success and explanatory power, Einstein's general theory of relativity was not immediately accepted by the scientific community. In fact, it took several decades for the theory to gain widespread recognition and experimental validation, and even today, there are still some unresolved controversies and open questions surrounding it.

One of the main reasons for the initial skepticism towards general relativity was its sheer mathematical complexity and abstraction. The theory was based on advanced concepts in differential geometry and tensor calculus, which were unfamiliar to most physicists at the time. Even Einstein himself struggled with the mathematical formalism, and relied heavily on the help of colleagues like the mathematician Marcel Grossmann to develop the theory.

Moreover, the predictions of general relativity were so counterintuitive and seemingly fantastical that many scientists had a hard time accepting them. The idea that space and time were not fixed and absolute, but rather were dynamical and interconnected, was a radical departure from the classical Newtonian worldview that had dominated physics for centuries.

Some of the most famous predictions of general relativity, such as the bending of light by massive objects and the existence of black holes, were initially met with skepticism and even ridicule by some members of the scientific community. It was not until the 1919 solar eclipse expedition, led by the British astronomer Arthur Eddington, that the bending of light was finally confirmed experimentally, providing the first direct evidence for the theory.

Even then, however, there were still some lingering doubts and controversies surrounding general relativity. Some scientists argued that the theory was too speculative and untestable, and that it relied too heavily on abstract mathematical concepts that had no physical meaning.

Others pointed out that there were still some unresolved issues and inconsistencies within the theory, such as the problem of singularities (points of infinite density and curvature) and the incompatibility between general relativity and quantum mechanics. These issues remain active areas of research to this day, and have led to the development of new theories and models that attempt to reconcile the two pillars of modern physics.

Despite these challenges and controversies, however, general relativity gradually gained acceptance and recognition within the scientific community. This was due in large part to the accumulation of experimental evidence that supported the theory's predictions, as well as the growing recognition of its explanatory power and elegance.

One of the most important early tests of general relativity was the precession of the perihelion of Mercury, which had been a longstanding problem in Newtonian mechanics. Einstein's theory was able to explain this anomaly perfectly, without the need for any ad hoc assumptions or modifications.

Other important tests of general relativity included the gravitational redshift of light (the stretching of light waves as they climb out of a gravitational field), the Shapiro time delay (the slowing down of light signals as they pass near massive objects), and the Lense-Thirring effect (the dragging of space-time by rotating massive objects). All of these effects have been confirmed experimentally to a high degree of precision, providing strong evidence for the validity of Einstein's theory.

In the decades since its initial publication, general relativity has become one of the cornerstone theories of modern physics, alongside quantum mechanics and the Standard Model of particle physics. It has led to a profound transformation in our understanding of the nature of space, time, and gravity, and has opened up entirely new fields of research and discovery, from cosmology and astrophysics to gravitational wave astronomy and quantum gravity.

Today, general relativity is widely accepted as one of the most successful and well-tested scientific theories in history, with a wide range of practical applications and technological spinoffs, from GPS navigation to gravitational wave detection. It is a testament to the incredible power of human curiosity and creativity, and a reminder of the profound beauty and complexity of the universe we inhabit.

At the same time, however, there are still many open questions and unresolved issues surrounding general relativity, particularly when it comes to its compatibility with quantum mechanics and its implications for the ultimate fate of the universe. These are active areas of research that continue to push the boundaries of our understanding, and that will undoubtedly lead to new discoveries and insights in the years and decades to come.

A Final Word

Dear Readers:
Thank you for reading this book. Please write a review, and share with your friends on social media if you enjoyed this title. As a small independent publisher, we use our resources mainly for the research and development of high-quality books that we can bring to you, and do not spend significant resources on marketing and advertising. We are counting on **you** to spread the word!

Dear teachers, librarians, and educators:
Do consider the books in this series for your class activities and exercises. Our titles are also available on leading databases like Overdrive, Scribd, BorrowBox and Hoopla.

Explore more titles from Lexicon Labs in the pages that follow.

Follow Dr. Leo Lexicon on X for news updates and promotions @LeoLexicon

Explore the lives of great innovators, Scientists, Leaders, Artists and Explorers...Stay tuned for additional titles coming soon!

**Learn the basics of Coding and program in Python.
No prior knowledge required!**

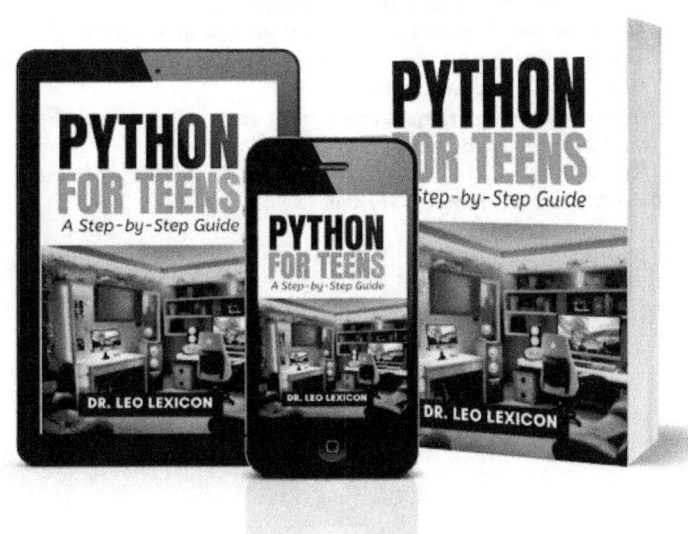

Meet our bestselling titles on AI
BOOKS FOR CURIOUS MINDS

- Structured introduction to the building blocks of AI
- Review of major milestones in AI history
- Meet the leading inventors and their key innovations
- AI concepts explained in a simple, easy-to-understand format by a Bay Area educator
- Resources for puzzles, games, and coding
- Perfect travel companion or gift

Follow Dr. Leo Lexicon on Twitter/X

 𝕏 @LeoLexicon

LEXICON LABS

LEARN ALL ABOUT STARTING AND GROWING A BUSINESS AS A TEENAGER

Explore the Future of Quantum Computing

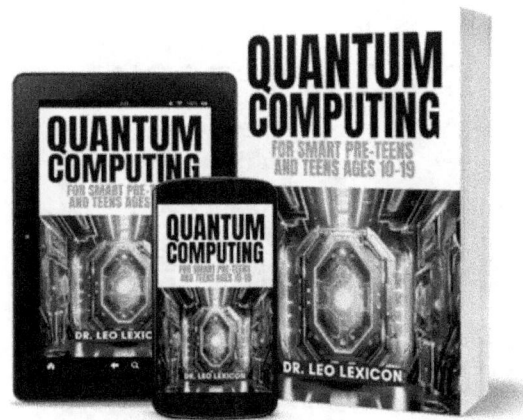

FUN BOOKS FOR TRIVIA NIGHT

COLORING BOOKS

TEST YOUR INNER NERD!

Discover More Bestselling Titles from Lexicon Labs!

SCAN ME

Education, Entertainment, and inspiration.
GUARANTEED.